The Enigmatic Expanse Existence

By

Sreedhar Iyer

Copyright © 2018 Sreedhar Iyer

All rights reserved.

ISBN: 9781718198807
ISBN-13: 9781718198807

CONTENTS

1	Prologue	1
2	Religious beliefs	9
3	Spiritualism	15
4	Scientific view on the creation	19
5	Introspection	55
6	Free will	71
7	The pale blue dot	85
8	Conclusion	89

Prologue

The bus started off from beach city Karwar and made its way over the bridge that connected the two roads, which were otherwise separated by the river Kali. Under the bridge, the river formed a delta joining the Arabian Sea. On one side of the bridge, the view was picturesque with the river starting to spread out into the delta, and on the other side, there was the big sea. A couple of ships were seen near the port. Three weeks prior I had booked a five-day camp-stay at Dandeli resorts. The camping schedule included trekking, river-rafting, bird-watching and various other activities in the deciduous forests of the Western Ghats of India.

I had suitably settled into a window seat enabling me to enjoy the natural scenery. I immediately fell in love with what lay around me. The beautiful scenery was in stark contrast to the city life which was filled with lifeless buildings, roads, never-ending traffic and the highly polluted air. But here the ambience was different; it was serene, with water gently flowing from the river as it joined the great Arabian Sea. The entire horizon was surrounded by a range of mountains. The air was fresh and clean. It was the third week of August and the rain had subdued after a continuous downpour of over two and half months.

In general, the rains of the Western Ghats are copious enough to keep ample water flowing in the rivers and tributaries for the rest of the year. Even months after the monsoon ends, the waterfalls around here, small and large, never seem to run out of water.

As the bus rolled over the hilly terrains and zigzagged along the mountain edges, I watched in awe the exquisite scenery that surrounded me. There were lush green trees and mountains on one side and the river on the other side. Villages and agricultural fields alternately appeared between stretches of dense jungle. The morning sunlight reflected from the crop circles enthralled me and the cool breeze from the wetlands grazed my skin rejuvenating me. I hurriedly dug into my bag seeking my camera, still undecided if I had wanted to capture the scenery or just enjoy the beauty that lay before me.

For every few kilometers, villages having a small population appeared in-between dense forests. It made me wonder whether the villagers here were safe from the wild animals that would chance to roam around. It was later clarified by my escort Mr. Sachin Kamath that, even though wild animals like bison, deer, porcupines, leopards and sometimes tigers were seen, they rarely had inflicted any violence against human beings. Other animals such as fox, hyena, mongoose, the great Malabar squirrels, snakes and other reptiles too ruled these forests. I had glimpsed peacocks flying from one tree to another, making it an exciting trip. After a two-hour drive, the bus reached my destination. On de-boarding, another escort guided me to the resort. He suggested that I freshen-up before we began our trek. Excited as I was, I took no time to get ready. Armed with my digital SLR and a backpack I was ready to embrace Mother Nature.

Our trekking began on a well-established trekking trail along with a few other nature enthusiasts. Birds such as Bulbul and Kingfishers were abundant and easily evident. Migratory birds and other enigmatic birds of different species that were present in the dense trees needed special attention to spot them. The birds moved so rapidly that it was difficult to capture them through the lens of my camera. Endless trees and shrubs spilled everywhere. Had anyone been tasked to carry out a census of the trees and shrubs, it would be herculean one. It would be similar to counting the sands on the seashore. Walking on the trail, we had to cross over small streams that crossed our path.

Mr. Sachin cautioned us to watch out for leeches that may silently creep up on our legs and suck blood. In wet areas, I saw large populations of worms. They were non-existent during my previous visit last summer. They seemed to have come into existence after the onset of monsoon. Each time, after the first rain of the year, loads of worms and peculiar insects of various shapes and sizes suddenly appear in the forests. Thousands of frogs suddenly appear from nowhere and croak near water-pools which are created by rain. During nights, insects such as crickets and flying ants would suddenly enter through window cracks attracted by artificial sources of light at nearby residences. Wet surfaces during incessant rains invite growth of algae and moss on almost all surfaces. I wonder where all these creatures and organisms were during the rest of the year, and how did they appear suddenly? It would seem that they waiting to be germinated, or, were they hiding in an unknown location somewhere deep in the forests only to come out during the rains?

Earth is a mysterious place and the only planet known to host life. It is a wonder that we are born in such a unique planet that has a well-balanced ecology which supports co-existence of plant and animal life. The trees provide food and oxygen, the essential survival kit, to the human beings and the animals. Waste of the animals combines with soil to produce manure for the plants. The nitrogen in the atmosphere helps the plants produce necessary survival components like amino acids.

Surely we must have been created by a creator or else how does one explain such a wonderful place that keeps us alive and suitable to us in every way. But one big question crops up in our minds is - Why? Why were we created and why should we exist at all? What is the purpose of this life? We go through a cycle of birth, growth, and death. The legacy continues with the birth of our offspring and their cycle starts. This process which started several thousands of years ago is still continuing. What is the meaning of this? **Why do we simply exist?**

The perfect environment

Our Earth is a marvelous place to be in. We live in an environment that has a tolerable temperature, humidity, pressure and other parameters which make our life comfortable. The abundance of food makes us habitable on this planet making it a unique place in the whole of the cosmos. Neither have we heard of, nor have we seen any another place that possesses this magic. There is no evidence of life on any other planets in our solar system or any other solar systems as far as we can see. Millions and millions of plants and animal species survive on this planet. What makes the Earth such an apt and beautiful place to live?

A wonderful planet positioned perfectly from the sun so that we are neither baked nor frozen to any extreme. When compared to the other planets of our solar system, our surroundings are truly beautiful and the environment is comfortable for our existence and continuation of life. Physicists call the position of our planet in the solar system as the Goldilocks zone. It is a band of the region in the solar system where it is neither too hot nor too cold for water to exist in liquid state. Also, the Earth has a day-and-night cycle so that we can do our business, dine and have fun during the day while we can rest during the night. We have our own sun which provides us the energy to stay alive, and there is the moon that shows itself during the nights. The moon can be seen during the day too, but we seldom notice it on a bright sky.

Earth is the third planet from the sun and is located at around 150 Million Kilometers away from the sun. The first two planets are Mercury and Venus. Mercury being very close to the sun at 57.9 Million Kilometers revolves around the sun in 88 days but takes a whopping 59 days to make a complete rotation around its own axis, causing it to be tidally locked with the sun. One side of the planet is baked to blazing 800 degrees F while the other side is super-cooled at minus 290 degrees F. These two extreme temperatures are so harsh that it is impossible to imagine life on planet Mercury.

The second planet Venus is a barren planet and is known to have a suffocating carbon dioxide atmosphere which is 90 times thicker than Earth's. It contains clouds of sulfuric acid and extremely hot surface temperatures that reach 864 degrees F (462 degrees C). Leave alone trying to live there, it may be even difficult to just visit this planet.

The fourth planet is Mars, whose atmosphere is about 100 times thinner than that of the Earth's. Without a "thermal blanket," it cannot retain any heat energy. On an average, the temperature on the planet Mars is about minus 80 degrees F (minus 60 degrees C). A summer day on Mars may get up to 70 degrees F (20 degrees C) near the equator, but at night the temperature can plummet to about minus 100 degrees F (minus 73 C). Even though research is on to identify if human colonization is possible on Mars, it needs some more time and experimentation to identify the hardships that such a colony will face.

The outer four planets: Jupiter, Saturn, Uranus, and Neptune are super-large gas giants. They do not have any rocky surface on which we can stand on. Their atmospheric pressures of these planets are so enormous that one would be crushed immediately on entering the atmosphere.

Our visibility outside the solar system is not good enough to see if there are any civilizations present in the neighborhood. It is only assumed that intelligent species, if any, will try to communicate through radio waves just like we do, and hence governments of developed countries have devised programs such as "Search for Extraterrestrial Intelligence" (SETI) to scan the cosmos for any intelligent signals or messages that may give us some insight on any extraterrestrial civilizations which may be lurking around somewhere in the vast expanse.

Sreedhar Iyer

Religious beliefs

Religion is a system of faith that believes in a supernatural-being a.k.a. the God or the creator, and tries to establish a relationship between man and the God. Religion has played a very important role in honing the lifestyle of the majority of the human population. Religion has matured enough to control the moral ethics of the society in which it has been established. Geographical disconnection of the human population has resulted in the formation of different religions. The number of religions on our planet is about four thousand. Each religion has its own Gods and the people in these religions have their own method of appeasing their Gods. Each religion has its own rules and methodologies for rituals and religious procedures. Most of us are in one religion or the other by birth. Let's analyze what the major religions have to say about our existence?

The Christian belief of creation (Abrahamic):

Genesis is the creation myth of both Judaism and Christianity and its stories are found in the Book of Genesis. The seven days of creation mentioned in the book are as follows:

Day 1 - God created the heavens and the Earth. He further created light and separated the light from the darkness, calling light as "day" and darkness as "night."

Day 2 - God created an expanse to separate the waters and called it "sky."

Day 3 - God created the dry ground and gathered the waters, calling the dry ground "land," and the gathered water and called it the "seas." On day three, God also created vegetation (plants and trees).

Day 4 - God created the sun, the moon, and the stars to give light to the Earth, to govern and separate the day and the night. These would also serve as signs to mark seasons, days, and years.

Day 5 - God created every living creature of the seas and every winged bird, blessing them to multiply and fill the waters and the sky with life.

Day 6 - God created the animals to fill the Earth. On day six, God also created man and woman (Adam and Eve) in his own image to commune with him. He blessed them and gave them every creature and the whole Earth to rule over, care for, and cultivate.

Day 7 - God had finished his work of creation and so he rested on the seventh day, blessing it and making it holy.

Many other religions believe in a similar creation theory where importance is given to Earth, sky, stars and human beings.

The Hindu belief of creation (Indian):

The Hindu tradition perceives the existence as cyclical nature of the universe and everything within it. The cosmos follows one cycle within a framework of cycles. It may have been created and reached an end, but it represents only one turn in the perpetual "wheel of time", which revolves infinitely through successive cycles of creation and destruction. Of the various Puranas or the sacred texts of the Hindus, it is Brahmanda Puraana which tells us about creation from the eternal entity (or the Brahman) which neither has a beginning nor an end. It adapts formula of the Pañca-lakṣaṇa, pronounced as Pancha Laxana - meaning five characteristics. These are:-

(I) Sarga (Creation of the Universe),

(II) Pratisarga (Dissolution and re-creation),

(III) Manvantaras (Periods of Time presided over by Manus),

(IV) Vaṁśa (Genealogies of Gods, the Patriarchs, the Sun and the Moon) and

(V) Vaṁśānucarita (accounts of dynasties of different ruling families).

(I) Sarga

In the beginning, the entire universe is dark (unmanifested) as the Gunas (characteristics) are in Laya (equilibrium). Only the Brahman or the ultimate reality exists. The Gunas are Rajas (passion and activity), Sattva (goodness, positivity, truth, balance, virtuousness) and Tamas (destruction, chaos). The disturbance in the Gunas results in the manifestation of a self-born deity - Brahma who has three functional states; Brahma with Rajas, Vishnu with Sattva and Rudra with Tamas. Brahma creates the universe, Vishnu protects and Rudra destroys, and then the cycle repeats.

Primary Creation- These are the Prākṛta (created from Prakṛti) :

(1) Creation of Mahat-Tattva - The basic material energy. From this Mahat-Tattva, the intelligence or buddhi manifests along with the sense of 'Aham' or Ego.

(2) Tanmātras - Creation of air, sound, fire, water, and Earth.
(3) Vaikārika - Creation of Aindriya Sarga or sensory perception.

Secondary Creation-
(4) Mukhya Sarga -Creation of immovable things like mountains, trees etc.
(5) Tiryak-Srotas - Creation of lower creatures like birds and animals.
(6) Ūrddhva Srotas - Creation of Gods and divine beings.
(7) Arvāksrotas - Creation of human beings.
(8) Anugraha-Sarga - Creation of other species with magical powers are created such as the Yakshas, Gandharvas, Sarpas, Apsaras etc. Some of these species are in the mode of goodness, having Sāttvika Guna while others in the mode of darkness having Tāmasika Guna.

Secondary creations (4-8) are called Vaikrta and these function without consciousness. The sacred sound Aum is believed to be the first sound at the start of creation.

(II) Pratisarga – Dissolution, and re-creation of the universe
(1) <u>Nitya Pralaya</u> - Every day - the death of people.
(2) <u>Naimittika Pralaya</u> (Periodical) - takes place at the completion of a Kalpa which is one day of Brahma or about 4,320,000,000 (4.32 Billion) human years. A Kalpa equals 1000 Maha Yugas where each Maha Yuga consists of four yugas; namely, Satya Yuga, Treta Yuga, Dwapara Yuga and Kali Yuga. In accordance with Hindu belief, we are in Kali Yuga

which started around 3102 BCE. On completion of Kalpa, the Samvartaka fire burns down the four worlds namely, Bhū, Bhuvar, Svar, and Mahar. Thereafter, Samvartaka clouds pour down torrential rains and everything, mobile and immobile, is destroyed and dissolved into one vast expanse of water and then God Brahma goes to sleep for a period of one Kalpa. After this Brahma wakes up to recreates everything.

(3) <u>Prākrtika Pralaya</u> - takes place at the end of Brahma's period -about 311 Trillion 40 Billion Human Years or 100 Brahma years. When the withdrawal of the Universe is imminent, at the completion of the life-duration of Brahma, the Earth-matter will be resolved into water; the Water into heat; heat into air; air into ether; the ether into essential matter (Bhutādi), the essential matter into the principle of intellection; the principle of intellection into Prakriti (Nature); and Nature into Purusha (the eternal subjectivity) and finally only the Atman remains. This process of re-absorption of Principles (Tattva-Sāmya) is also recurrent. The universe will remain so for another 100 Brahma years, after which a new Brahma is created and a new cycle starts again.

(4) <u>Atyantika</u> -Ātyantika Pralaya takes place when one becomes liberated through spiritual knowledge. He does not take up another body just as a sprout (never comes out) when the seed is burnt. This is called "dissolution due to dawning of knowledge".

Sreedhar Iyer

Spiritualism

A Spiritualist is a person who believes that we are all made up of soul or energy and are part of the bigger soul which some consider as supreme soul. The concept of spiritualism varies and has differences geographically and some have religious backing. Thus spirituality is being taught by different Gurus, and the method of teaching varies.

Western spiritualists believe that the soul, which is in a different realm, enters the body during birth and goes back to the spiritual realm at the time of death. It is believed that departed souls can interact with the living. Spiritualists claim to make contact with the dead, usually, through the assistance of a medium. A medium is a person believed to have the ability to contact spirits directly.

Islam doesn't view spirituality separately from everyday activities. In Islam, everything is spiritual because all actions must be in accordance with God's pleasure. This view comes from the Islamic creed and the Muslim's understanding of tawhid (the oneness of God). According to Islam, God has appointed the human soul as His Khalifah (vicegerent) in this world. He has invested it with a certain authority and given it certain responsibilities and obligations for the fulfillment of which, he has endowed it with the best and most suitable physical frame. The body has been created with the sole object of allowing the soul to use it in the exercise of its authority and the fulfillment of its duties and responsibilities. The body is not a prison for the soul, but its workshop or factory; and if the soul is to grow and develop, it is only through this workshop. Consequently, this world is not a place of punishment in which the human soul, unfortunately, finds itself, but a field in which God has sent it to work and do its duty towards Him. In Islam, spiritual development is synonymous with nearness to God. Similarly, he will not be able to get near to God if he is lazy and disobedient. In Islam, the distance from God signifies the spiritual fall and decay of man.

Hinduism has no centralized religious authorities, no governing body, no prophets nor any binding holy book. Hindus can choose to be followers of multiple deities or identify God with the universe or can be monistic, which means that there exists only a single thing which can only be artificially and arbitrarily divided into many things. A Hindu can also be a non-believer in God. Within this diffuse and open structure, spirituality in Hindu philosophy is an individual experience and referred to as ksetrajna. It defines spiritual practice as one's awareness of self, the discovery of higher truths, true nature of reality, and a consciousness that is liberated and content. It is a journey towards moksha (salvation) by the way of merging into the supreme soul. Hindu spiritualists believe that human beings primarily crave for materialistic goals to achieve satisfaction and happiness. However, after achieving these goals, they are still unhappy and begin to chase new dreams.

Superlative and long-lasting happiness can be attained through spiritual practice. Spirituality has to be experienced and cannot be experienced by reading books. When we do spiritual practice without expectation of worldly gain and with sincerity, then God takes care of our material needs and let our spiritual practice continue. Hindus believe that when the soul departs from the body, the life-breath follows. When the life-breath departs, all the organs follow. Then the soul becomes endowed with particularized consciousness and goes to the body which is related to that consciousness. It is followed by its knowledge, works, and past experience. When a soul assumes a human body, it takes up the thread of spiritual evolution of its previous human birth and continues to evolve toward Self-knowledge.

According to Hinduism, all souls will ultimately attain self-knowledge. For those who attain self-knowledge before or at the time of death, their souls become absorbed in Brahman (or the supreme soul or the universe).

Sreedhar Iyer

Scientific View on the creation

Let us analyze what science hypothesizes about the creation of the universe and life. But before we dwell into the theory of creation, we'll have a brief understanding of basic science and the quantum physics.

Almost everyone would have learned science as a subject in the school. Science is a systematic process that builds and organizes knowledge in the form of testable, repeatable explanations and is based on testable theories. A new theory is accepted only if it succeeds in explaining a phenomenon for which, there was no explanation earlier and, at the same time explains other related phenomena for which there may be an existing accepted theory. When the new theory is accepted, the old theory is dumped. This process is a continuous one, and it may be pertinent to note that, a sound and logical theory or hypothesis which holds good today, may totally be replaced by another one in the future. It must also be understood that, although science tries to explain all the phenomena occurring in this universe, it has still not made a breakthrough in many.

The branches of science that we all know since our elementary school are Physics, Chemistry, and Biology. Physics deal with forces and physical properties of matter. Chemistry is a branch of science that deals with shell-level bonding, formation of compounds and molecules; and interactions between matters chemically. Biology deals with anatomy and processes, in living organisms like plants and animals.

Building blocks of matter

Atoms are said to be the basic building blocks of all the matter in the universe, be it living things or non-living things. Let's try to understand what these atoms are. Atom was described as the smallest division of natural matter, where no further division was further possible i.e. chemically. The concept of an atom being the smallest indivisible matter was of philosophical nature until 200 around years ago. Indian sage and philosopher Kanada came up with the idea that anu (atom) was an indestructible particle of matter, while Greek philosopher Democritus in the year 400 BC was the first person to use the term atom (atomos, meaning indivisible). But physicists later discovered that atoms are not the fundamental particles of nature but are conglomerates of even smaller particles. In the 1890s, Physicist J. J. Thompson discovered the electron, which was a further sub-division of an atom, with the help of a cathode ray tube.

In 1910, a series of experiments conducted by Earnest Rutherford led to the revelation that the atoms were largely empty having a positively charged dense nucleus at its center. This discovery paved the way for the discovery of the proton. In the year 1932, James Chadwick discovered the neutron which gave a final picture of the atom. A picture of an atom, that was similar to the planetary model, having a central nucleus consisting of protons and neutrons, around which the electrons revolved, in a fashion very similar to the planets revolving around the sun. The electrons had negligible mass and had a negative charge, whereas the nucleus possessed most of the atom's mass.

Planetary Model of Atom

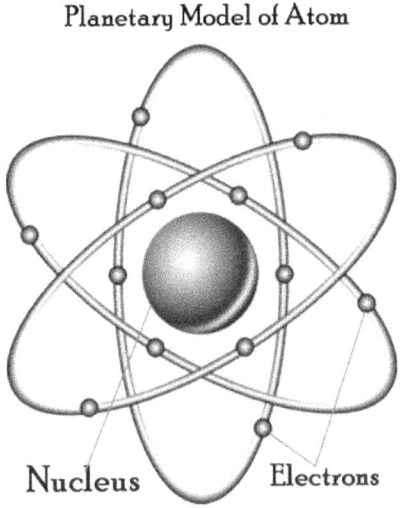

Nucleus   Electrons

I remember my school physics text depicting an atom with a central positively charged nucleus consisting of protons and neutrons and electrons revolving around the nucleus, just like planets revolve around the sun. We neither had the questioning attitude nor did we manage to ask our teacher why they behaved this way. It was the age of indoctrination and we just were indoctrinated. Cool! Those few who did not continue in the science stream, still believe the atom to be in this configuration.

When the planetary model of the atom was proposed, there were questions as to why the electrons, despite having an opposite charge to that of the protons, did not fall into the nucleus due to electrostatic attraction similar to pieces of paper being attracted to a statically charged surface. Some theories were proposed using "classical physics" model to explain why the electron did not fall into the nucleus, but the explanations remain unsatisfactory. Newtonian mechanics, thermodynamics, and Maxwell's theory of electromagnetism are all examples of classical physics. It was seen that many theories in classical physics break down when applied to extremely small objects such as atoms or to objects moving near the speed of light. This gave birth to "quantum physics" which explained the behavior of subatomic particles.

Wavering logic – The Wave-particle duality:

The wave-particle duality is a phenomenon which baffles anyone who has read about or encountered it. We know that light travels from sources like the sun and stars to our planet, thus making them visible to us. We know that the sound is a form of energy and it travels through various mediums such as air, glass, metals, etc as waves. The presence of the medium assists the flow of sound energy. It has been observed that sound travels up to a few thousands of miles, whereas light travels from the far reaches of outer space. Light has to pass through a vast extent of empty space wherein the matter is sparse. So what exactly is light and how does it travel such great lengths? Is light a particle or a wave?

There were two different views as to what light was: In the early 16th century, Christian Huygens and Isaac Newton proposed opposing theories of light and its behavior. Huygens proposed that light must be a wave, while Newton proposed that light is a stream of particles (corpuscles). Huygens believed that light was a longitudinal wave and that this wave was propagated through a material called the 'aether'. Since light can pass through a vacuum and travels rapidly Huygens had to propose some rather strange properties for the 'aether': for example; it must fill all space and be weightless and invisible.

Corpuscular theorists described light corpuscles as single, infinitesimally small particles which have shape, size, color, and other physical properties.

In 1800s, a breakthrough experiment by Thomas Young which proved that the light indeed behaved like a wave shook the claim of corpuscular theorists. In the famous double slit experiment, Thomas Young passed sunlight through a double-slit cut-out from cardboard and directed the resultant image on the screen behind the slits. The image on the screen was expected to be two parallel lines, but instead, it showed a fringe pattern. (See fig-2)

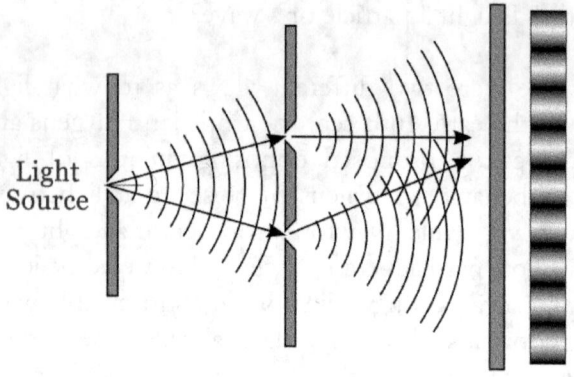

A fringe pattern is a pattern of dark and bright lines, whose appearance is only possible if light behaved like waves. When more than one wave meets, they create troughs or crests. Wherever the waves add up, a crest is formed with a bright line, and wherever the waves cancel out each other, a trough is formed creating a dark line. A similar phenomenon can be observed when stones are thrown in still water.

Later in 1905 when Einstein postulated the special theory of relativity he attributed light to be of dual nature having both particle and wave property.

If Thomas Young's wave-particle duality experiment came as a surprising evidence to prove that light behaved like a wave, the experiment by Claus Jönsson in 1961 came as a bolt from the blue.

Claus Jönsson, a German physicist carried out the double slit experiment using an electron beam and a recipient fluorescent screen. The aim of the experiment was to observe where the electrons hit the screen after it passed through the double slit. The result of the experiment came as a surprise, as it was observed that the pattern created on the fluorescent screen was similar to the one created by the light passing through the double slit.

The experiment was repeated again, but with a variation. This time, the electrons were released one at a time. The result was no different, which simply meant that the individual electrons did not go through one slit but went through both the slits creating an interference pattern. This baffled the whole scientific community. How could electrons, which had a mass and considered to be matter, manage to go through both the slits at the same time? The only explanation suggested was that the electrons too switched its state between wave and particle. This experiment did not end here. Further experiments showed wave-particle duality was also observed with atoms and larger molecules such as fullerene molecule which consisted of 60 carbon atoms. The question that bothered us was, how, particles such as electrons, atoms, or molecules could pass through two slits at a time and manage to interfere with themselves. Attempts were made to identify from which slit the particles passed, but when such an attempt was made, the interference pattern disappeared and the results matched with that of particles fired at the double slits; only two lines were seen on the screen.

Initially, it seemed as if nature did not like us peeping into its activities beyond a certain point, but later it has been understood that, by trying to measure the position of the particle near the slit, we are actually disturbing the process that occurs near the double slit. The double slit experiment has revealed that matter and energy are analogous and exist in dual nature.

Fundamental particles:

Scientists found that the smallest conceived particles that were sub-atomic, such as the protons and neutrons were themselves not the smallest particles, but were formed by smaller entities which are now defined as elementary particles. These particles are covered under particle physics or quantum mechanics. Elementary particles are the smallest constituents of something and aren't made up of even smaller particles (not yet). Particle physics addresses both matter and force in terms of particles, so elementary particles come in two types.

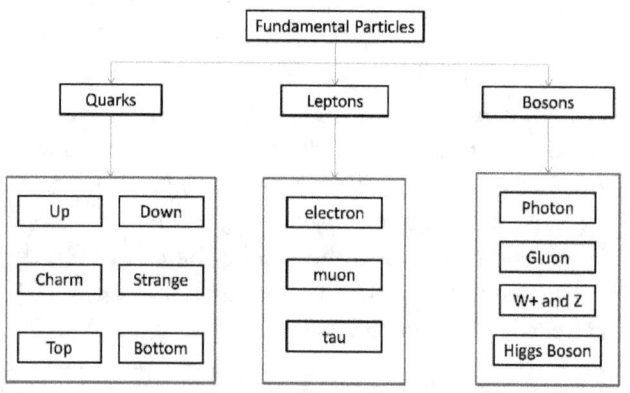

The elementary particles are broadly categorized as Bosons and Fermions. Bosons are force-carriers and the Fermions come together under the influence of other Fermions as well as bosons (Gluons) to make up the matter.

There are six flavors of quarks: Up, Down, Strange, Charm, Top, and Bottom.

Up and Down quarks have the lowest masses of all quarks and they are generally the most stable and most common in the universe. Protons are made of two Up and one Down quark. Neutrons are made of two Down and one Up quark. Leptons are elementary particles that do not undergo strong interactions. Electrons fall under the category of charged leptons. The other types of leptons are muon and tau which are very unstable and decay very quickly.

The diameter of an atom is calculated to be around 1 to 5 angstrom (1 angstrom $=10^{-10}$m), whereas the diameter of a nucleus is around 2 fm (1 femtometer = 10^{-15} m) 15 fm. Hence it can be seen that the area occupied by an atom is a million times larger than that of its nucleus. In the vast space between them, the electrons are distributed and are understood to be discreet wave functions having fixed energy levels. Now, this was a brief wrap-up of the basic science and quantum physics. Now let's see what science has to say about creation.

The Big Bang Theory

Of the many theories doing rounds in the science fraternity, a few decades back, it was hypothesized that our universe formed with a big bang approximately 13.8 billion years ago. The Big Bang theory is an effort by physicists to explain what happened at the very beginning of our universe. According to this theory, our universe could have sprung into existence as "singularity"; an infinitesimally small, infinitely hot, infinitely dense singularity. With the tremendous amount of energy in the form of heat, initially, only leptons, quarks along with gluons, W and Z bosons and photons would have existed. The high temperatures would not have allowed the formation of protons or neutrons.

With the help of modeling, scientists have deduced that during the initial period, an equal amount of anti-matter would have existed, but due to reasons unknown, matter has prevailed giving rise to all the objects we see today in the universe. The modeling also indicates that during the initial seconds the universe would have expanded and cooled rapidly, thereby allowing the formation of protons and neutrons.

Within the first 300 seconds of the existence of the universe, the elements helium, lithium, heavy hydrogen (deuterium) and Helium-3 would have formed from the protons and neutrons by a process called nucleosynthesis. For the first 380,000 years after the Big Bang, the intense heat from the universe's creation would have made it essentially too hot for light to shine. Atoms would have crashed together with enough force to break up into dense, opaque plasma of protons, neutrons, and electrons that scattered light like fog. 380,000 years after the Big Bang, with the universe expanding, matter would have cooled enough for electrons to combine with nuclei to form neutral atoms. This phase, known as "recombination," would have enabled the absorption of free electrons by the atoms enabling the universe to have become transparent. The light produced during this time is detectable even today, in the form of radiation from the cosmic microwave background, albeit the difference is that the light waves have expanded towards the microwave range since we are receiving the light from 13.8 billion years ago. It is to be noted that this is the same in all the directions. The universe would have expanded like dots on a balloon with the dots also being stretched out.

400 million years after big bang, clouds of hydrogen gases would have started to collapse due to gravity, becoming hot enough to start a nuclear fusion and form stars. In the star's core, the nuclei of hydrogen are understood to have combined to form helium. With the passage of time and due to extreme pressure, other elements of higher order have been created by the fusion of nuclei. The initial stars were of population-III type of stars which were big red giants and did not have metals. Current theoretical stellar models show that most high-mass Population-III stars quickly exhausted their fuel and exploded in extremely energetic pair-instability supernovae. Such explosions are understood to have thoroughly dispersed their material, ejecting metals into the

interstellar medium further enabling them to be incorporated into the later generations of stars that were population-II and population-I stars. Large volumes of such star debris would have collapsed to form galaxies. Gravitational attraction pulls galaxies towards each other to form groups, clusters, and superclusters.

Our sun belongs to the later generations of stars and its formation is estimated to have started around 8.5 to 9 billion years after the big bang. It too seems to have incorporated debris of the previous generations of stars as seen from the ingredients of our solar system. Even though the big bang theory tries to explain the process of the universe formation 13.8 billion years ago, it does not explain what exactly triggered the big bang.

Evidence for the Theory

1. The redshift of galaxies:

Redshift is a phenomenon which occurs when light from an object has an increase in its wavelength when the light emitting object is moving away from the observer. Conversely, it may also be said that the light wave shifts to the red end of the spectrum. Redshift is an example of the Doppler Effect which is apparent in the pitch of sirens and frequency of the sound waves emitted by speeding vehicles.

Upon observing the sky, it is seen that the galaxies appear to be moving away from us at speeds proportional to their distance, meaning: the farther they are, the faster they are moving away. This phenomenon is called "Hubble's Law," named after Edwin Hubble who discovered this phenomenon in 1929. This observation supports the expansion of the universe and suggests that the universe was once compact.

2. Microwave Background:

Secondly, if the universe was initially extremely hot as the Big Bang suggests, we should be able to find some remnant of this heat. In 1965, Radio astronomers Arno Penzias and Robert Wilson discovered Cosmic Microwave Background radiation (CMB) which pervades the entire observable universe. This is thought to be the remnant which scientists were looking for. Penzias and Wilson shared in the 1978 Nobel Prize for Physics for their discovery.

3. Elements found:

Finally, the abundance of the "light elements" Hydrogen and Helium found in the observable universe are thought to support the Big Bang model. Sun, our own star contains heavier elements compared to far-away stars which indicate that earlier stars were made of lighter elements.

Big Bang Theory – Is it the Only Plausible Theory?

Is the standard Big Bang theory the only model consistent with the available evidence?

No, it's just the most popular one. Internationally renowned Astrophysicist George F. R. Ellis explains, "People need to be aware that there is a range of models that could explain the observations. For instance, I can construct you a spherically symmetrical universe with Earth at its center, and you cannot disprove it based on observations. You can only exclude it on philosophical grounds. In my view, there is absolutely nothing wrong with that. What I want to bring into the open is the fact that we are using philosophical criteria in choosing our models. A lot of cosmologists try to hide that."

In 2003, Physicist Robert Gentry proposed an attractive alternative to the standard theory, an alternative which also accounts for the evidence listed above. Dr. Gentry claims that the standard Big Bang model is founded upon a faulty paradigm (the Friedmann-Lemaitre expanding-spacetime) which he claims is inconsistent with the empirical data. He chooses instead, to base his model on Einstein's static-space-time paradigm which he claims is the "genuine cosmic Rosetta."

Dr. Gentry has published several papers outlining what he considers to be serious flaws in the standard Big Bang model. Other high-profile dissenters include Nobel laureate Dr. Hannes Alfvén, Professor

Geoffrey Burbidge, Dr. Halton Arp, and the renowned British astronomer Sir Fred Hoyle, who is accredited with first coining the term "The Big Bang" during a BBC radio broadcast in 1950.

String theory:

Apart from the big bang theory, there are various other theories that are making rounds. For instance, we have the string theory which hypothesizes that all matter is made up of tiny vibrating strings. In String Theory, the electron is considered as a tiny string vibrating at a particular frequency while a proton is considered as another string vibrating at a different frequency. In this manner, what we envisage as fundamental particles in classical physics, are explained as minute strings vibrating at its resonance frequency.

While the big bang theory tries to explain the phenomenon of the universe from a starting point, the string theory hypothesizes what could have happened before the big bang. The basic idea of the string theory is that elementary particles are not point-like, but rather infinitely thin one-dimensional strings. The large zoo of elementary particles, each with its own characteristic properties, reflects the many possible vibration patterns of a string. When the strings encounter other physical entities, called "branes", they open up into line segments. The interactions of these strings and branes would then produce the various forces and reactions we see in physics. While normal physics views the universe in four dimensions (three for space and one for time), various versions of String Theory assume that the universe has as many as 11 dimensions. These assumptions are attempts to unify all of the known properties of physics into a single "theory of everything."

String Theory, in all its forms, is purely theoretical. No part of String Theory has ever been demonstrated, tested, or observed in any way that could confirm or disprove it.

There is another theory (incidentally a part of string theory) which hypothesize that we may be living in a 2-dimensional hologram. It suggests that our universe might be two-dimensional, except, we perceive it as three-dimensional. The flat surface of the universe contains all the information we need to sense three dimensions, much like a hologram. String theory says that the gravity in the universe is made up of thin, vibrating strings called gravitons. These strings make up the holograms of events that happen in 3D space within a flat cosmos. They contain the information that a 3D object would have, for example, its volume. So if you want to know what's happening inside a ball, you can find out everything you need to know just by looking at the surface. This is only a thought experiment and, in accordance with physicists, it is a reasonably acceptable theory that works on the mathematical model.

According to another theory suggested by professors from Oxford University and Massachusetts Institute of Technology, the universe may be a giant supercomputer and we may all be living in a simulation, possibly in a simulation of some game played by someone who is much more complex than we can even imagine. Imagine being the Mario in the game Super Mario Bros, which is a 2-dimensional game developed by Nintendo Entertainment System. Nintendo has 8 lives. He lives in a mushroom kingdom and must rescue the princess after fighting with the minions. It is quite possible that we may be living in a simulation that allows us to see some things and keeps some things hidden or unexplainable.

Are we just a set of information?

The Earth rotates around its axis and revolves around the sun. The Sun along with our whole solar system orbits around the center of the Milky Way Galaxy. Galaxies are moving apart at great speeds.

Are we the same person that we were yesterday? Cells in our body pass on and get supplanted by new cells persistently. Red blood cells can live for approximately four months, while white blood cells can live more than a year. Skin cells have a life of two to three weeks. Colon cells live for about four days. Up to 10% of cells in skeleton get supplanted in a year. The number of cells that an adult male loses per minute is roughly 96 million. At the same time, approximately, 96 million cells divide and replace these dead cells. Just as we shed dead skin cells, dead cells from internal organs pass through and out of the body with waste products.

The food we eat and the air we breathe helps the replacement process of the cell tissues. Information of the old the cell is copied to the new one which takes its place. Thus it is seen that after a few years, we would be completely replaced except for neurons in the brain and lens cells in the eyes which never get replaced. It is amazing that we still keep the same identity and similarity of the person who previously existed. The only addition is that we have some additional memories of experiences in-between and mutations due to change in environment.

Living and nonliving things:

Research indicates that all living and non-living things are made up of atoms. To date, scientists have discovered around 118 elements. The elements start with Hydrogen which has one proton, and one electron having an atomic number of 1. The next is Helium with two protons, two neutrons and two electrons, and it has an atomic number of 2. The list goes on till we reach the element number 118.

Two or more atoms of the same or different elements, bond chemically to form new compounds which have totally different characteristics from their parent elements. For example, carbon forms allotropes of graphite or diamond, both having drastically different physical and chemical properties. Oxygen having atomic number 8 combines with two Hydrogen atoms having atomic number 1 to form H_2O. Both Hydrogen and Oxygen are gases; hydrogen being highly explosive and oxygen a supports combustion, but when they chemically bond, it forms water which is neither explosive nor is a fuel to fire and exists in the liquid state.

Nature creates a balance from imbalance, by converting elements into molecules. These molecules may bombard among themselves physically or react chemically with other elements and molecules. A chemical reaction may require external energy, or it may produce additional energy during the conversion process. Sometimes room temperature is sufficient to trigger chemical reactions between the molecules or elements. For example, iron combines with water at room temperature to form iron oxide (rust) and hydrogen.

Nature creates molecules, but do they have a life? But before we dwell into that, let us understand what we mean by life. First and foremost, we consider ourselves as living beings and that we have something called life. But what separates us from the lifeless things around us?

There is no clear definition as to what life is, but to chalk out a few let's start with the basic things we can gather. Our life has a beginning and an end, thus we are born and so we die. But this happens with other things too such as stars. Between the process of birth and death, living things grow, but so does stars. We consume energy by ingesting food and expelling the waste. Again we know that stars convert hydrogen to helium and use up the energy to keep it going. It expels heat and light. We respire as we need oxygen to convert the glucose absorbed by cells into energy. Our lungs deliver oxygen to the blood, and in return takes back excess carbon dioxide back into the atmosphere. The excess carbon dioxide that we animals produce is absorbed by plants and trees, which in turn gives oxygen thereby completing an ecological balance sustaining both plants and animals.

The most important aspect of life is reproduction. Another aspect is that living things can move. But we also know that the stars and star systems evolve, stay active for a few billion years and explode, and the process of birth and death of star system continue. The population of stars started from zero and now is running into trillions. In a real sense, they too are reproducing.

But life, in a sense, is different. Living beings have a process of life and death, are sensitive to the surrounding environment, reproduce to keep its species continuing, consumes energy in one form to sustain and throws out the waste in a different form, which is used by the ecology. In addition living beings have emotional states that are complex and intangible. The emotional states include feeling of hunger, anger, love, compassion, jealousy, empathy, happiness, sorrow and so on, which are helpful for self-sustenance in a community-based system.

Spiritualists see living things as soul and the body. They argue that body is materialistic and lifeless. It is only the soul that gives the necessary life to the body. Scientists are unsure how the living things came around in a very vast non-living materialistic universe and have proposed a number of theories trying to explain the presence of living things in the universe.

Digging deep into the anatomy of living plants and

animals, it is observed that all living things are made up of tissues and organs. All the tissues and organs are made up of a fundamental unit called the **cell**. A cell is the smallest unit of life that can replicate independently and they are often called the "building blocks of life". It has been observed that cells of the smallest bacteria to the largest of whales, carry-out similar functions and also have a similar structure.

The cell theory states that:
- All living organisms are made up of one or more cells.
- All the life functions of an organism occur within cells and,
- All cells come from preexisting cells.

Scientists estimate that our bodies roughly contain 75 to 100 trillion cells. Cells do everything, from providing structure and stability, provides energy and means of reproduction for the organisms.

More about cells

<u>Prokaryotic</u> single-celled organisms were the earliest and most primitive forms of life on Earth. Examples of prokaryotes are bacteria and archaea.

Animals, plants, and fungi are examples of organisms that are made up of <u>Eukaryotic cells</u>.

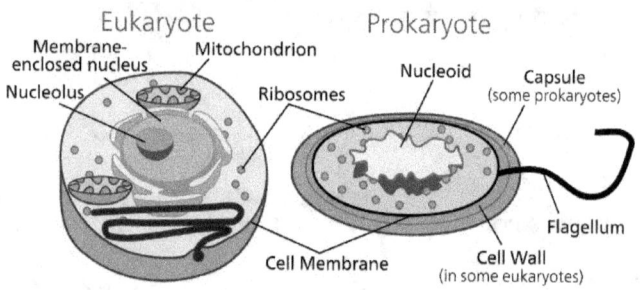

Cells contain DNA (DeoxyriboNucleic Acid) and RNA (RiboNucleic Acid), which are complex molecules known as nucleic acids. They carry the necessary genetic information for directing cellular activities.

In prokaryotic cells, the single bacterial DNA molecule is not separated from the rest of the cell but coiled up in a region of the cytoplasm called the nucleoid region. In eukaryotic cells, DNA molecules are located within the cell's nucleus.

Cells contain structures called organelles which have responsibilities that include everything from providing energy to producing hormones and enzymes.

Mitochondria in the cell are basically responsible for providing cellular energy. Additionally for signaling, cellular differentiation, cell death, as well as maintaining control of the cell cycle and cell growth

The cell membrane is a thin coat of lipids that surround a cell. The cell membrane is made up of millions of smaller molecules that create a flexible and porous container. It is primarily responsible for containing the cell structure. Secondly, it has to allow certain type molecules in, and a certain type of molecules out, through its pores.

In the 1950's, biochemists Stanley Miller and Harold Urey conducted an experiment which demonstrated that several organic compounds could be formed spontaneously by simulating the conditions of Earth's early atmosphere. They designed an apparatus which held a mix of gases similar to those found in Earth's early atmosphere over a pool of water, representing Earth's early ocean. Electrodes delivered an electric current, simulating lightning, into the gas-filled chamber. After allowing the experiment to run for one week, they analyzed the contents of the liquid pool. They found that several organic amino acids had formed spontaneously from inorganic raw materials. These molecules collected together in the pool of water to form coacervates.

The experiment used water (H_2O), methane (CH_4), ammonia (NH_3), and hydrogen (H_2). The chemicals were all sealed inside a sterile 5-liter glass flask connected to a 500 ml flask half-full of liquid water. The liquid water in the smaller flask was heated to induce evaporation, and the water vapor was allowed to enter the larger flask. Continuous electrical sparks were fired between the electrodes to simulate lightning in the water vapor and gaseous mixture, and then the simulated atmosphere was cooled again so that the water condensed and trickled into a U-shaped trap at the bottom of the apparatus. After a day, the solution collected at the trap had turned pink in color. Using paper chromatography, Miller identified five amino acids present in the solution: glycine, α-alanine, and β-alanine were positively identified.

In a 1996 interview, Stanley Miller recollected his lifelong experiments following his original work and stated, "Just turning on the spark in a basic pre-biotic experiment will yield 11 out of 20 amino acids."

The original experiment remains today under the care of Miller and Urey's former student Jeffrey Bada, a professor at the UCSD, Scripps Institution of Oceanography. The apparatus used to conduct the experiment is on display at the Denver Museum of Nature and Science. Their experiment, along with considerable geological, biological, and chemical evidence lends support to the theory that the first life forms arose spontaneously through naturally occurring chemical reactions. However, there are still many skeptics of this theory who remain unconvinced.

In March 2015, NASA scientists reported that, for the first time, complex DNA and RNA organic compounds of life have been formed in the laboratory under outer space conditions, using starting chemicals, such as pyrimidine, found in meteorites.

A short history of life on Earth:

By observing a variety of star systems, physicists have worked out the way our solar system may have formed. Around 4.5 billion years ago, our Solar system along with our nascent Earth is assumed to have formed from the swirling dust and gas remnants of an old star's supernova explosion. Chunks of matter and gas grouped to form the sun and the planets that we see today. The nascent Earth's atmosphere is said to have been composed of carbon dioxide, water vapor and nitrogen. The formation of our Moon is attributed to a possible collision with another planet soon after Earth's formation. It is thought that warm oceans gradually formed, from steam escaping from the crust, volcanic activity and collision by icy meteorites, soon after the Earth's formation.

Studies indicate that early Earth faced harsh environmental conditions which included constant bombardment by asteroids and enormous volcanic activity. During this period it would have been highly hazardous for any life to exist, but all the necessary ingredients were present in some form or another.

Liquid water and chemical building blocks; namely: oxygen, hydrogen, carbon, nitrogen, sulfur, and phosphorus were present. It is not known what exactly triggered the process of life on our planet. Whether it was comets which brought in the seeds of life from other civilizations in outer space or a natural event or process that might have been triggered in such environmental conditions, life began slowly, starting with single-celled organisms and bloomed into millions of species with different traits. In the initial days, there thrived single-celled organism known as Cyanobacteria (also known as blue-green algae) and it is one of the earliest types of prokaryotic bacteria. Its fossilized remains have been found in Australia dating back to between 3.4 and 2.8 billion years ago. These fossil remains indicate that the organisms were already biologically complex, with cell walls protecting their protein-producing DNA, suggesting that life actually began much earlier, perhaps as early as 3.8 billion years ago. These early Cyanobacteria were the first oxygen-producing organisms. They evolved continuously and were phototropic i.e. positive response to sunlight. They were responsible for the initial oxygenation of the Earth's atmosphere, as they produced oxygen while sequestering carbon dioxide in organic molecules during the period from 2.7 to 2.2 billion years ago. In this process, they produced organic carbon, the building blocks of life's molecules, and released oxygen gas (O_2). The rapid explosion of life began after oxygen became abundant.

The first eukaryotic-celled organisms having one or more complex cells evolved sometime between 2.5 and 1.7 billion years ago, perhaps coincident with the rise in atmospheric oxygen during that period. Each cell contained a nucleus which was surrounded by a membrane that held the cell's genetic material, holding and protecting the complex molecules such as RNA and DNA.

The first multi-cellular life probably arose around 1.2 billion years ago, in geological terms almost overnight, while the landmass of the Earth was still a single continent called Rodinia. It presumably started out as a sort of symbiosis, a loose cooperation between single cells that gradually became more and more complex. Recent research on single-celled organisms called choanoflagellates has yielded the rather surprising fact. It has been deduced that single-celled organisms began communicating with each other and effectively working together as a single unit.

Sexual reproduction evolved, leading to faster evolution where genes are mixed in every generation enabling greater variation for subsequent selection. 600 million years ago it is thought that the earliest multicellular animal kicked off with sponge-like creatures. This was followed by mutations that led to the evolution of cnidarians which possessed nerves and muscles.

Evolution appeared to speed up again about 550 million years ago with the sudden appearance of the first hard-bodied animals which possessed brains and eyes. The Cambrian explosion which lasted for about the next 20 million years resulted in the divergence of most modern metazoan phyla and the diversity of life began to resemble that of today.

The first vertebrates emerged 500 million years ago followed by fishes. 390 million years ago some fresh water lobe-finned fish developed legs and gave rise to the Tetrapoda. 360 million years ago the amphibians evolved where the animal kingdom took its first steps to walk on the land.

300 million years ago the reptiles evolved and laid its eggs on the land. 256 million years ago shortly after the appearance of the first reptiles, two branches split off. One branch is the Sauropsids making up the dinosaurs, modern reptiles, and birds. The other branch is Synapsida, from which come modern mammals.

85 to 65 million years ago mammals (rodents, rabbits, and pikas) evolved which lived on lands and resided in lower branches of trees. A possible meteor strike wiped out large reptiles like dinosaurs. The extinction cleared the way for the mammals, which went on to dominate the planet. 63 million years ago was the age of monkeys and apes, and after this, the journey of evolution continued to Hominids. Homo sapiens evolved only 160,000 years ago.

Between 60,000 and 12,000 years ago humans migrated out from Africa, to Australia, India, and Europe. Approximately 23,000 years ago humans began cultivation of land for agriculture. The industrial revolution started around 250 years ago and the internet revolution began a few decades ago.

Variations in design:

Look around and notice that no two objects seem to be exactly identical. In this vast ocean of matter, we would expect some match here and there, but we see that nothing seems to be exactly identical. There is no structure that has an identical pair. We would expect the products of a mass producing factory to be identical, but each product is different although they are manufactured using the same process and raw materials.

A handful of sand from the beach seems to have similar shape and size. But when magnified by around 200-300 times under a microscope, it reveals that they are widely varied and there are no two grains closely similar. Not only the shape, but the color, translucence and other properties are starkly different making it difficult to find two sand grains that are identical.

Sand magnified under a microscope (Image source: sandgrains.com)

Even at the quantum level, scientists hypothesize that no two electrons can have the same four sets of quantum numbers, thereby creating a variance in subatomic level. Considering the diversity of non-living matter, say our own solar system; we observe that the planets in our solar system too are diverse, even though they were formed from a single source of gas and cosmic dust. It is observed that none of the planets are in any way similar to the rest of the planets, either in composition, spin or any other property. We see life blooming only on the Earth and know of no other planets that host life.

On Earth, there are infinite trees in the wild but none are alike, each different from the rest. We find that the leaves on a tree too are different from one another, even though they are from the same roots. Indeed this is the same everywhere. This is the way nature seems to work. These variations seem to be the law of the universe and they seem to be random in nature. We see that in the higher order creations like humans or animals, this variation is clearly visible. We observe that one person is a dud, whereas the other one is intelligent. One is handsome whereas the other one is not. One is quiet whereas the other one is talkative. With all these traits we find people with multiple combinations of characteristics that make them unique. The human population on the Earth has touched seven and half billion and yet we observe that each of us is unique. There is a saying, that there are six other persons in the world who resemble you, but eventually when we find one such person we notice that the similarity is limited to physical appearance alone. The only eventuality of striking similarity is found in the case of identical twins, where appearances are deceiving, but there are bound to be subtle differences in appearances or in their personalities that can tell them apart.

Although the offspring or descendants that sprout or evolve resemble their parents, they are not a ditto copy. Variations are seen in the newer generations that may be minor to major. Minor variations include differences in their heights in comparison to the parents. Major deviations too are observed when the newer generation does not resemble its parents and seem to have very different properties than its parents.

It may be seen that all the pups of the same litter have different colors or patterns. One or more pup will look like it does not belong to the family at all. This phenomenon is not only with dogs and cats but can be observed in almost all the species including humans. You may see this type of variations in your family too. Nature does not seem to want all pawns on its chess boards to be the same. If all the pawns would have been similar, they would all have carried out similar functions, but again nature does not want this arrangement.

Since the day the universe began, progressive change has been its mantra. Initially, it was the conversion of energy into matter and anti-matter. Then it was the evolution of fundamental particles, which was followed by its merger into subatomic particles and then the formation of atoms. Later with abundance in elements, it was time for the evolution of compounds and molecules and then to highly complex molecules called the biological cells. Simple biological cells mutated to multi-cellular organisms continuing its journey to the evolution of mammals. The universe has been evolving and mutating, creating millions of biological species which ensures the sustenance of the ecology that it has nourished. Nature has always been using the principle of variation or mutation in the production of offspring. This has enabled it to create offspring with different qualities, which in turn increases the probability of the species' survival against predatory threats and severe climatic conditions. There may have been short-necked and long-necked giraffes, but the long-necked ones survived. Green grasshoppers have a color camouflage to avoid being noticed by its predator. Nature has embedded diversity in the process of its journey from the big bang. It is designed to carry out an experiment of deviations, for, if there were no deviations or variations, life would not have been possible, and we would never have evolved.

Sometimes I wonder if I had been the creator of the universe, what would I have done differently. I would have started off by creating colorful worlds having uniformly decorated skies. In these worlds I would have created mangroves having sweet mangoes, orchards having beautiful apples, trees providing other fruits, and fields of corn which produced healthy cereals that never perished. I would neither have thought of creating weeds and unwanted trees nor would I have thought of creating insects like cockroaches or mosquitoes that keep me on the edge. But the creation is not bothered about such organized scenarios that would appease the human beings. It just keeps on churning stars, galaxies, species of animals and plants and odd things that we are yet to encounter. Further, it creates variations in each of the species so that the best one matching the environment, adapts and survives, thus creating a sort of continuity in whatever it is trying to produce. It seems as if nature or the universe is in process of a big experiment. It seems to be doing a copy and paste with definite in-built variations so that the next product will be definitely a different one. Why and how nature does this is a big question.

Survival probabilities:

An aspect that keeps me pondering is the concept of the pollination and the mechanism of pollen's dispersal and distribution. Pollen grains are produced in the stamen of flowers. These are the minute particles that look like powder and contain the male reproductive part of the plant. By a process called pollination, the pollen grain attaches itself to the carpel, which is the female reproductive part of the plant to form the seed. If the pollination occurs in the same plant, it is called self-pollination. If the pollen grains are transferred from one plant to another, to geographically different locations by wind or migratory birds or insects, it is called cross-pollination. It is a wonder, that nature utilizes various methods to increase the cross-pollination by multiple methodologies which helps in creating diversity. We notice that nature seems to diversify the genetics of the seedlings through random permutations and combinations of cross-breeding. But is it really random and does the nature following definite rules? We have learned from physics and mathematics that all particles follow a definite set of rules and nature cannot break the universal rules.

The blueprint:

Look at the seed thus formed, it is so small but contains all the information required to grow into a full plant or tree. What information does the seed possess that enables it to convert into a big tree when it encounters suitable environmental conditions? Millions of seeds dispersed by plants and trees lie dormant until the first raindrop awaken them. They begin to sprout when the monsoon sets in and the time is ripe for them. The tiny seed encapsulated in a hard shell contains the entire blueprint for it to grow into a fully developed plant or tree if situations are favorable, and nature ensures that the situations are favorable through various means.

Intelligence and awareness

Before the evolution of humans, it may seem that nature has been carrying out the process of evolving, replicating and mutating, thus creating stars, planets, and complex organisms. But somewhere down the line, human beings evolved possessing capabilities of self-awareness, the ability to analyze and make judgments. It seems as though the Universe created beings (us) that could think about themselves and try to reason their own existence.

Now! This raises a good lot of questions. How did this suddenly happen? Sudden; when we consider the age of our cosmos. Billions of years in evolving species, that only supported ecological cycle of the planet, and suddenly there is a spurt of intelligence. Now, what is the intelligence that we are speaking of? Why are we aware of our own presence in this universe? Are we the only ones to be self-aware or, are there some things that we are yet to know?

Well, intelligence gave humans the power of forethought, methodology to communicate, gather into communities, start civilizations, cultivate land, and implement cultures and traditions, and develop technological implements that try to mimic the nature or even try to surpass it. In a short period of sixty thousand years, the species on our planet changed from ignorant to self-aware. If it is true that Earth hosted life for the first time in the universe, then it can also be said that the universe has become self-aware within a minuscule span of its total life cycle. But the question is; Are we the first?

Sreedhar Iyer

Introspection

I was born in a cosmopolitan city. I really cannot remember all the activities from my birth. But I do remember some fragments of my childhood days; only some glimpses and some fragments. I remember the streets through which I walked to my school during my childhood days. I remember going to my school along with my cousins and friends. It is all stored in my memory like a summary. The human brain is an amazing thing. Unlike a video recorder which we can play back in sequence, the brain stores long-term information in an overlapped manner. A detailed event is stored in case of specific events; say a very joyous event or in case of a trauma or distress. I remember that during my childhood, I used to commute to my school using a well-defined route, but sometimes I used to take other routes too. My memory remains smudged, as to when I took the main route and when I took the alternate routes.

I remember a few of my teachers, but I do not remember many. I remember my vacations and joyous encounters with my cousins that happened once in a few years. I remember getting engaged to my spouse and the day of my wedding. The routine thereafter is stored as a summary. I recollect the birth of my children, and their progress is only stored as a summary.

During my younger years, I have been introduced to religious practices like offering prayers and chanting hymns and have been guided by my parents to be devout.

"Pray to God and you will excel in your exams", my mother used to tell me. I would abide by her and pray to God, to get a good score on my exams. Little did I know that I needed to study hard too, for God to help me get a good score? Results, of course, were not the best, no matter how much effort I tried to put in my prayers.

"Be patient." my mother used to tell me. "Keep faith in him and everything will be alright". Being Hindu, we have ample count of Gods and Goddesses. We have a Goddess for knowledge, and a Goddess of wealth, a God for removing obstacles and so on. Moreover, we had no restriction on praying to the Gods of other religions. I had prayed to all the Gods during my difficult times, but the results were always not as expected. Finally, I lost trust in the God and stopped my prayers. Bingo! There was no change and my life continued as usual. It seemed as though God was not worried about losing a devotee. Life continued as usual but I felt that there was something amiss the way we see God. There were too many religions, too many Gods, and too many procedures that took our time.

One day, I asked myself "Does God really exist? And if he did exist, then did he listen to our prayers? Moreover, if I and my rival pray together, whom will he support?"

These questions put me in a new thought process. I reasoned out three logical possibilities.

1. God exists, and listens to our prayers, but he tests us to the extreme, to identify if we really believe him. There is no human logic that can understand his whims.

2. God exists but does not have any intention to listen to each and every living being on the Earth. He lets natural events to take over the process that goes on.

3. God simply does not exist. Everything was created by accident.

In the first case, God is like what everyone imagines, but he/she is testing you. Mythology is flooded with stories indicating that very few devotees have succeeded in contacting or connecting with the God after long periods of penance. Does this mean that I was only doing immature things, just waiting for good results and still not knowing how far away I'm from connecting with him or from success?

The second case is possible if God exists as we think, but either doesn't have a method to connect with us or, he/she is not inclined to hear all the fuss we make about things small and big.

The third case is hard to believe logically as it is difficult to comprehend how something could be created without a creator, but the same question arises, in case it is assumed that God exists and created the world around (i.e. who created God?).

With lots of theories about God and creation doing rounds, we still are very far away from answering questions on our existence and our purpose on the Earth.

Everyone around you has an explanation as to why something did not go right in your life. The priest advises you to pray God more fervently and to continue keeping faith in him; the statistician says that you had missed it only by a chance; the spiritual guru says that you have to 'let it go'.

We are born, we grow, we reproduce and we die. The cycle repeats itself with our children going through a similar process that we had undergone, but with what benefit? Our children start everything from scratch, just as we did. Lessons that our parents had learned, remains with them and do not get transferred to us, and our experiences do not pass on to the next generation. Hence it is seen that nature does not have the plan to transfer our experiences to the next generation. The reason is not obvious, but it may be due to the way our nature is structured. If we were able to pass on our experience and awareness to the new generation then we would be leapfrogging in terms of technology and in solving the mysteries of universe and creation. But currently, our efforts remain in teaching the next generation about what we learned in our generation, and by the time the new generation reach our level of understanding, they are ready to transfer their experiences to the next generation. Ultimately, generation to generation we are making very little headway in understanding the reason for our existence and that of our surroundings.

It is a human tendency to be egoistic, selfish and

arrogant. Since birth, we feel that everything around belongs to thyself. Sharing resources and compassion are imbibed in us while we grow up in a society. But as individuals, we have the mental feeling of being mighty, extraordinarily strong and very highly intelligent. But when such feelings are to be implemented in real-world scenarios, the extraordinary powers seem to vanish and we become a normal being.

Long ago, humans were trapped on the planet due to the belief that the Earth is the center of the universe. After the revelation that the Earth is only a tiny speck in a humongous expanse, we still are trapped on the planet due to gravity and other life supporting dependencies. We are aware of the universe because we see the stars in the night sky. With new discoveries like telescopes, we can peer furthermore into space. We are limited due to our dependence on the nature around us and due to the fact that we have not built sophisticated equipment to explore the universe at a faster pace. But even with these limitations, we feel that we are the most supreme and most intelligent beings in the Earth. But are we? What are the qualities that make us feel supreme beings on the Earth?

After the supremacy of Dinosaurs and their untimely death 66 million years ago, evolution has taken many turns. In the current twenty-first century, it is the humans who rule this planet. What makes us supreme or what makes us feel that we are more powerful compared to other species on the Earth?

Some reasons are cited below:

<u>We are large in numbers</u>:

The first and the foremost advantage that we have over the other species is that we are large in number. We occupy almost all the land mass. Our footprints exist in urban, rural, and dense forests too.

<u>Communication and culture</u>:

Since ages known, we humans have learned to express our intentions, ideas, and experiences to other members by way of speech, sign or writing. By communicating with each other we can convey our ideas to others and transfer knowledge between one another, thereby taking our collective knowledge to higher peaks. Communication between the people is mostly verbal and is based on a variety of languages and dialects which vary according to geographic location. Cultural programs like folklore, dancing, singing, drinking, and feasting are in vogue from time immemorial. Recent innovations in technology have brought communication to home and further to handheld devices by use of wireless technology. With the advent of the internet, technology has connected the whole world, which has far-reaching consequences for the future of mankind.

<u>Reasoning ability</u>:

An important ability that humans have is the ability to reason, judge and take decisions for events that are yet to take place. This ability helps us in carrying out calculated planning and achieving tasks that are otherwise only probabilistic; few examples are driving a car, launching a satellite into space, marketing, exploration etc.

This ability has enabled humans to undertake various research and development projects, and this has helped us further in understanding about various hidden secrets of the world around. Reasoning and research have led to developments in technological implements enabling us to reach new frontiers that would have been otherwise impossible.

Imagination:

I am not aware of any animal that has the capability to imagine, but to us human beings, it is a regular and never-ending activity. We dream during our sleep, day-dream when awake and we constantly imagine. Sometimes our imaginations are normal but many times they are wild and preposterous. It is said that imagination is the key mantra for development. There are both positive and negative aspects of imagination. When we imagine about scenarios which are within our reach or reasonably achievable, then progress is seen, but if the imagination is wild or, it is based on fantasies, then it becomes wastage of resource and time.

Technology:

The human empire mutated from hominids family around 250 thousand years ago. The human race initially lived in caves and forests eating leaves, fruits, and nuts. Later they learned the art of hunting by making tools. Tools were initially made out of stones, and later with iron. Early Humans discovered the fire and utilized it for cooking. They settled near river beds, cultivated lands, and produced crops for consumption. Then they grouped together to form communities giving rise to civilization. When we compare humans with other species on Earth, we can say that we have developed exponentially in terms of technological implements. It may have started off with the invention of hunting implements and agricultural implements, but the Industrial revolution, internet revolution, and various other scientific innovations have put the humans technologically ahead within a very short span of time, especially in the fields of machinery, transportation, communication, space exploration, robotics, and artificial intelligence. Currently, we live in a modern civilization with capabilities to communicate throughout the globe, technology to navigate the entire Earth, travel to the outer space and possibly colonize other planets. It can be said that other living beings living on Earth are nowhere near us in terms of technological advances.

Our limitations:

I don't know what to call this setup we live in. Whether it be called world, nature, universe or multi-verse. Our perception of the world around us is a combination of what we see, feel, touch, smell and hear. All the information collected by our sense organs is collected and compiled by our brain, which gives us a picture of the world around us. Among the sense organs of our body, taste and touch are contact-based, whereas smell and hearing work within small distances. These four senses help us in our survival in the immediate environment that we live in. The eyes are a gift to us which facilitates us to see the world and helps us to navigate around, read, write and help us to do umpteen other chores in our day to day activities. Moreover, it has helped humanity in looking outside our planet too.

We primarily receive light from the sun, our own star. We need light to distinguish between objects. If light does not exist then we cannot see anything. Our eyes are useless in the dark. We do not see light directly, but we see objects when they are exposed to light. Our eyes have rods and cones which receive reflected light from objects, convert it into electrochemical signals and then send it to the brain. We have two eyes separated by a few millimeters horizontally, and each eye individually senses two-dimensional images (namely, the breadth and the height). Our brain combines the signals from both the eyes and creates a three-dimensional perception which helps us navigate the terrains on Earth.

With our limited sensing capabilities, we cannot perceive if more dimensions exist, than we actually see or feel. Thus we can perceive only three spatial dimensions even if there are more dimensions as suggested by some physicists.

Our universe is very vast and contains trillion-trillion of stars, but we do not see all the light that is emitted by these stars Our visual ability is limited to visible light only, which include colors ranging through red, orange, yellow, green, blue, indigo and violet. We, humans, have been endowed with curious minds and are eager to know all about ourselves and the world around us, but our eyes cannot see everything.

Theorists explain the presence of an abundance of energy in form of electromagnetic radiation from stars and galaxies. The spectrum of electromagnetic waves includes radio waves, microwave, infrared radiation, visible light, ultraviolet radiation, x-ray and gamma rays. Visible light that we perceive is a range of electromagnetic radiation, which occupies only 0.0035% of the whole electromagnetic spectrum. Thus it may be seen that we are not able to perceive the remaining 99.996% of the electromagnetic radiation that exists in the universe. We get to know about the remaining energy with the help of special instruments and mathematical simulation. For this, we depend a lot on telescopes, spectroscopes, interferometers and various other instruments that can measure cosmic radiation and these signals are converted into a human-readable format for further analysis and interpretation. But this is only the tip of the iceberg. It is estimated that the stars and galaxies combined together form only 5% of the universe's total mass. The remaining 95% of the universe's mass is unseen and is presumed to be divided between dark matter and dark energy, which till date has not been observed by any means known to mankind. This reduces our visible perception of the universe to a meager 0.000175%.

Re-assessing the human senses, we see that our tongue being the taste receptor, when in contact with food items, helps us in identifying the edibility of the substance which we eat and thus can be categorized as an essential tool for survival. Nose, the olfactory help us identify whether the food we eat is suitable. The nose is also helpful in sensing hazardous environments. Our skin and hairs which are touch receptors help us feel around and help us in communication with dear ones. The eyes are helpful in navigating the terrains on the planet. Thus we see that the five basic senses we possess, simply exist to keep our survival on this planet a peaceful affair. A question that now crops up is, whether what we see and perceive about this world around is the same as what really exists.

Are other animals at par with us?

Are we the only ones who can communicate? We can observe tiny ants communicate among themselves making beautiful colonies. It's wonderful the way they carry their food. They leave a trail through which the other ants follow the path to reach its destination. Ants communicate with each other using pheromones, sounds, and touch. Pheromones are chemical substance produced and released into the environment by animals like insects and mammal affecting the behavior of others members of its species.

Chimpanzees greet each other by touching hands. Dogs stretch their front legs out in front of them and lower their bodies when they want to play. Dolphins are known to chat and they are unsurpassed in imitative abilities among non-human animals. Thus, it may be seen that animals have different methods of communication. They have been observed to communicate when they need to express to their mates about the location of food or about imminent dangers. With the vast species of animals on the planet, trials have been carried out only on few animals. As for the rest of them, it is not known whether they discuss, whisper or communicate in a completely different manner. Are there other animals having more sophisticated communication which may be better than our's, without us having an inkling of it? It is quite possible but presently unascertained. But it must be understood that not knowing about the prowess of other beings does not automatically make us superior in any way. It only emphasizes the fact that we do not have further information on the subject.

Further research is being carried out to study about the cognitive behavior and communication skills of non-human animals, but non-availability of proper methodology for such study has made the process a tad difficult.

Are we alone?

Physicists continuously seek an answer to the question that has been bothering mankind for a very long time. Are we alone in this universe? Or, is there anything or anyone out there in other parts of the universe who may have something in common with human beings, and, if they do exist, are they are trying to contact us? Such questions have been bothering mankind since the last 400 years.

Initially, the idea of alien civilizations is presumed to have started after the Copernican model of the heliocentric solar system was accepted while the concept of the Earth as the center of the universe was dumped. In 1899, Nikola Tesla observed radio signals that came from outside the Earth, which, he presumed to be a communication from other planets. This started off a debate among scientific communities on the possible existence of intelligence existing outside our Earth.

With the invention of the radio and its use in communication, scientists were of the impression that, an advanced alien civilization may try to communicate with others using the radio, just like we do.

Upon the suggestion by American physicist John D. Kraus, a flat-plane radio telescope named "Big Ear" was constructed in Ohio, in the year 1955, to scan the cosmos for natural radio signals. This was the beginning of the world's first continuous Search for Extraterrestrial Intelligence (SETI) program. In 1977 Jerry Ehman, a project volunteer of the SETI program witnessed a surprising strong signal from the space which was extraterrestrial in nature and seemed to be from an artificial source, but the signal was never received again.

Many countries have installed various radio frequency receivers to receive alien signals (if any), but even after several decades, no radio signals have been recorded that show evidence of intelligent life outside Earth.

We hear a lot of news about unidentified flying objects (UFO) sightings. We also hear of allegations on US government, accusing them of capturing aliens for carrying out research on them. But all these allegations are not yet proved. A large number of UFO sightings are reported in the USA, especially in California and Florida. Stories of aliens abducting humans for experimentation have also been reported.

Thousands and thousands of video footage on UFOs have been uploaded by citizens around the world and it is not known whether these videos are real, orchestrated or morphed. The followers of the alien theory are divided into two groups, one which believes that aliens exist and are hidden by the US government and the other group which fails to appreciate the trillions of video footages that show something sinister exists. It is difficult to verify the authenticity of these videos, and assigning the task to someone who will verify these videos with an unprejudiced view is bound to be ineffectual. It is also seen that some videos are shot in poor light.

A review carried out by government agencies revealed that 90 percent of the sightings could be easily attributed to astronomical and meteorological phenomena (e.g. bright planets and stars, meteors, auroras, ion clouds) or to such Earthly objects as aircraft, balloons, birds, and searchlights.

Review by psychologists on alien abduction indicated that sleep paralysis contributed largely to these amazing stories. Sleep paralysis is a disorder due to which a person is unable to move, either while awake or while falling asleep. During such an episode, one may hear, feel or see things that are not there and often results in fear. Such episodes generally last less than a couple of minutes. It may occur as a single episode or be recurrent.

With a lot of stories about UFOs, theories on aliens and alien abduction, there is no concrete evidence of any extraterrestrial life or intelligence.

Sreedhar Iyer

Free will

We know that all non-living things are made up of basic elements and molecules. The motion of bodies, such as planets, stars and other celestial objects can be predicted by knowing a few parameters such as their mass, angular velocity, and mass of nearby objects. The weather on the planet can be mathematically derived and predicted by measuring pressure, temperature and wind direction at various spots. By using scientific formulae, a physicist can predict the motion and interactions of particles in a closed jar, provided, the position and the physical properties of these particles are known. Thus it can be seen that the future events depend on the current state of various particles, their properties, and influence of nearby objects. Since all particles follow the laws of the universe which are definite, the eventuality is predictable.

If we consider the time from where the universe started, the forces that would have acted upon individual hydrogen atoms to group into clusters would have depended on the mass of the atoms and the neighboring atoms in its vicinity. Further accumulation of elements and sequence of process of formation of stars due to gravitational force would be within the ambit of the universal laws. Hence it can be said that each event would unfold depending on the previous state of the objects themselves and the influence of the objects nearby. Simply put, it means that the events that took place since the birth of the universe till date have followed a pattern and sequence, which depended on the initial condition similar to a story unfolding. The unfolding has been steady and definite, as there were no external stimuli (we do not know if our universe has anything outside it and it is assumed that there were no external stimuli). If all the parameters of each matter and their position in the universe are known at any instance of time, then predicting the next outcome is possible. This methodology is used by space scientists to launch spacecraft to faraway planets by using the gravitational tug of other planets to slingshot the craft to the required destination successfully.

That was about the predictability of non-living objects, but what about the fate of us living beings? Does it also follow the universal rule as it does for the non-living entities? Is our future predictable or do we have free will? Well, it is both as we see below:

Scientists say that living things, whether plant life or animal life, ranging from single-cell organisms having a size of around 0.1 micrometers (0.1×10^{-6} m) to large-bodied mammals and reptiles up to a dimension of around 30m length, are all made up of tissues which are made up of cells. Cells are made of proteins, water (a molecule) and other components which also are all made of molecules. Within the center of the cell are DNA and RNA, both extremely complicated molecules.

Categorizing by the types of elements, our body is made of only 11 types of elements which are: oxygen, carbon, hydrogen, nitrogen, calcium, phosphorus, potassium, sodium, chlorine, magnesium and sulfur. The different organs of our body are designed for specific functions. For example, the stomach is for digestion, lungs for absorbing oxygen from the atmosphere, the heart for distributing the oxygen to all the cells in the body. Tissues of the bones are made up of water, protein fibers and a mixture of mineral salts, calcium and phosphorous. The brain which is one of the most complicated organ in our body controls how the other organs of the body should function. The brain is the single point of processing and it has other roles too. It takes care of our voluntary and involuntary actions. Voluntary activities involve processes like walking, eating, drinking, etc. Involuntary activities include rhythmic and regular pumping by the heart, breathing, and metabolism. Metabolism refers to all the physical and chemical processes in the body that convert or use energy. It includes breathing, circulation of blood, controlling the body temperature, contraction of muscles, digestion of food and nutrients, elimination of waste through urine and feces, and functioning of the brain and nerves.

We feel hungry when the previous digestion cycle is complete. It is then that various organs send signals to the brain initiating the next round of ingestion. The brain tells the mind to find food, and the mind which is a virtual process of the brain works hard and helps us to find the food.

Moods and emotions are abstract features of our brain. Research indicates that moods and emotions that we experience are, in fact, release of certain chemicals in specific areas of our brain. At any given moment, dozens of chemical messengers or neurotransmitters are active, producing 'emotions'. Since the brain is our body's control center, such releases of chemicals in certain areas of our brain gives us the feeling that we are experiencing emotions such as happiness, sorrow, jealousy, greed, exuberance, etc. Emotions not only mean how we feel but also how we process and respond to those feelings.

The three most common emotion causing neurotransmitters are Dopamine, Serotonin, and Norepinephrine. Dopamine is related to experiences of pleasure and the reward-learning process. In other words, when we do something good, we're rewarded with dopamine and thus gain a pleasurable and happy feeling. This teaches our brain to desire to do it again and again. Serotonin is a neurotransmitter associated with memory, learning and easing depression. An imbalance in serotonin levels results in an increase in anger, anxiety, depression, and panic. Norepinephrine helps moderate our mood by controlling stress and anxiety. Recalling a negative memory can put us in a bad mood, and thinking about a happy memory can put us in a good mood.

For each of us, this process differs and it depends on the way various molecules are lined up and interact. Each of us has our own time for doing things, like waking up, taking food, or going to the library, etc. Let's try to analyze what exactly dictates that each of us differs.

Nature has created variations in each generation. Although we have the same sets of organs, our organ compositions differ physically. Each of us has a different brain; different, in the sense of configuration of the molecules in the setup of each brain. It may be the presence or absence of certain minerals, or, it may be a change in their sizes. These changes may be attributed to genetic reasons, climatic reasons and geographic reasons.

Each of us has different personal quality in our life. All the activities we do depend on this configuration which we normally say as "The way the individual's brain is wired". The reason we say "wired" is due to the fact that electrochemical signals are generated from the brain and sent to various organs for all the physical and mental activities.

Personal qualities include qualities such as sincerity, honesty, eccentricity, skepticism and so on. If each of these individual qualities were mapped on global level across the whole population, they would be following a normal probability curve; in the sense that, around 68% of the people will be in the center region, whereas the rest 32% will be divided into devoid or overzealous categories of the particular quality. A normal probability curve is shown below for illustrative purpose.

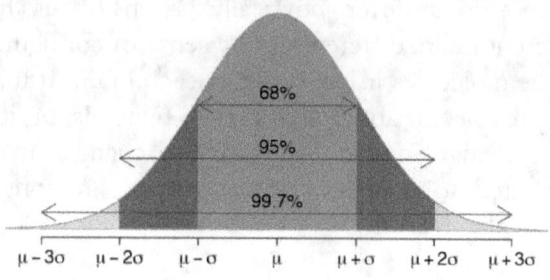

Assuming that we take one human quantity, say intelligence, it will be noticed that 68% of people will have average intelligence, 13.5% will be having below normal intelligence and 13.5% people will have above average intelligence, 2.5% on the lower side will be intellectually or disabled and 2.5% on the upper end will be extraordinarily intelligent or genius. This is for one of the human qualities. Similar to this there are more than a thousand qualities which can be tagged with human behavior and the distribution will be similar.

The combination of personal qualities with varying strengths makes up different individuals and affects the social setup. So when people mix around, you see people with different traits, dissimilar physique or varying mentality.

An interaction between two serious persons will be boring because both of their chemistries define them to be that way. Replace one person with a cheerful person, and then there will be a change in the interaction. So the eventuality of a meeting will depend on a person's brain setup and the person he interacts with (stimulant).

Since the whole of the human body is a molecular setup and the emotions we feel are said to be a chemical process, and since both molecular movement and chemical reactions are governed by universal laws; then by knowing the person's brain's chemistry and the stimulant, the next outcome can be determined. But such an exercise requires a very large scale computing, which is currently unknown. But assuming that in the future, we develop a powerful computing system which can observe the changes in the brains of the individuals and also assess the situation in the surroundings, then, we shall be able to predict the next outcome and thereby predict near future without any error.

To understand how the outcomes may fall out, let us consider two scenarios for the purpose of discussion.

Scenario one:

Edward, who is visiting Africa, is driving on a road that passes through a territory which is inhabited by a tribal group. All of a sudden a member of the tribe accidentally comes under the wheels of his car and is injured. The victim shouts in agony, thereby attracting other members of his tribe. They quickly surround Edward is quickly surrounded by the tribal people and are obviously angry and disturbed at the ghastly incident. On the molecular level, in their brains, neurons of specific areas of the brain are fired. But since each member of the tribe has a different brain setup, the intensity of anger and their curiosity about the incident differs.

Raaka, a member of the tribe who is among the angry crowd, is fierce by nature. He pulls Edward out of his car and proceeds to thrash him. Raaka has a personal quality to take up a challenge, and there had been a stimulus, a friend had been injured. The eventuality is that Edward is getting beaten. This new scenario is a stimulus for other tribe members who are with Raaka. A few of them feel that they have to join in and do their part, but there are a few others who would prefer to attend the injured. There are few other members who would want to caution Raaka to go slow for the fear that he may hurt Edward too badly. Whether they will stop the brawl or watch the scene purely depends on the structures of their brains. If their brain chemistry crosses the required threshold, they will stand between the two to mediate. Thus if we know the brain structures of all involved then the outcome will be predictable.

Scenario two:

In the first scenario, we see that the immediate future event depends on two factors. One is the present biological configuration of the persons involved and the second is the influencing processes, which further create a chain of events. The chain of events can be deduced if all the initial parameters before an event are known and all the laws of the nature are applied while deducing. But the next question that arises is, whether there is a chance or probability for an event like something out of blue, luck or charm?

Consider the following scenario. Assume that you are changing the channels on your television, and by chance, you stream a channel which at the particular moment is relaying a discourse by a renowned spiritual guru. Due to the sheer magnetism of the spiritual guru, you get deeply engrossed, and you immediately get to like the guru's preaching. Further, assuming that from that day onwards you decide to give up the Earthly pleasures and take up spiritual practice, then does this fall into the category of chance? That you accidentally switched on a spiritual channel and converted into a spiritualist.

Well, things get a bit complex now. In the first scenario, we had seen the emotional feelings of persons in distress. The outcome was predictable (certainly with knowledge of their brain chemistries, physical strength and a lot of computational skills). If the Raaka was not stopped, he would have severely injured Edward. It depended on how the others responded to the new stimulus. Were they timid? Or did someone gather their wits to stop the brawl? It all depended on their brain chemistries. In scenario one, the prediction was easier because events were local. But in scenario two, there seem to be a chance that you watched something that is not local, but was on television. Secondly, there was a randomness of you flipping the channels. Did it occur by chance? Was it predictable?

Let me explain to you why this too was predictable. You woke up at a particular time either depending on your body-clock-rhythm or due to a disturbance in the local environment (stimulus). It may be possible that your alarm woke you up or your spouse or, your mother had reminded you of the time. If your brain had overweighed your comforts over your commitments, you might have continued to laze in the bed. But actually, it was the other way around that you felt energetic and parted with your sleep. Your actions are in constant control of your brain, either by itself when you are alone or responding to a stimulus from the environment around you.

You walked into the living room because your brain told you to do so. You then switched on the television and then started to flick the channels. It was a routine and your brain was programmed to do it. Each channel had a schedule; hence there was no probability here. Assuming that when you had switched on the television, it had beamed your favorite soap opera that you had already viewed during the previous night, but you felt that it was not interesting anymore, so you had changed the channels until something interested you. What had decided this? Your brain chemistry decided this.

The person on the television was speaking in a soft tone. He was talking about an issue that had been bothering you from the past few days. He was providing ideas to overcome your problems. Curiosity had got better of you and you glued on to the channel. Your brain chemistry responded to this stimulus. Your mind (a brain's process) being vulnerable got convinced by his words and you made a resolve to follow the remedies that he suggested.

What I mean by vulnerable is that, if instead of you, someone with a diverse view had used the same remote, he might have skipped the channel and moved to Hollywood flick. But since your brain chemistry was molded in such a way that it tried to synchronize with whatever matched your frequency, and so you got fixated on the spiritual channel. So here too the result would have been predictable and there was no chance for 'a chance'.

All the processes in our universe whether macro or micro, small or big, in non-living entities or living beings, follow definite set of rules, no deviation whatever seen. As the size, location, birth and death of stars are predictable, so are the shape and position of the plants. Similarly the wind speed, rainfall, growth of the crops, involvement of the farmers and middlemen, the sale at the local supermarket, our selection of food, the method we follow to cook and eat, the way it changes the chemistry of our body including our brain and the next step we are going to take, are predictable. All are inter-related and predictable when the initial conditions are known.

If we had a similar universe elsewhere with the same laws of nature and the same beginning, then there should be no doubt that similar events would be occurring there on a similar Earth as it is here.

Does this mean that we do not have free will? Yes, surely freewill exists, but not the way we think. When we wish to do something by free will, it means that our brain is already wired to do it and our virtual senses have accepted it as our own will. The rest depends on the other minds that will be involved intentionally or unintentionally in addition to the processes involved, which we call as circumstances.

Your wish to wake up, to go to your school or work, to consume food surely will work, but your wish to get the best grades in examinations, perform well in your interviews and achieving your promotions depend on a lot of other factors such as, how your boss perceives your effort and sincerity or how tough the question paper is, who your competitors are, and so on. Although you may feel that it is your free will, it is actually the processes governed by the universe as each and every process are interconnected and interdependent.

Does this mean that our fate is written, our destiny sealed?

To answer this question, let us consider the way our universe evolved in around 13.8 billion years. It started with the creation of matter and antimatter from energy, which then annihilated or canceled themselves. But due to reasons unknown, some matter in form of fundamental building blocks remained. Further, these building blocks combined and decayed to form protons, neutrons and electrons and they prevailed. These subatomic particles further combined to form hydrogen atoms. From here gravity took reigns, stars formed and collapsed for many cycles, until heavier atoms sustainable for life were formed. Life on Earth started with single-celled organisms which reproduced by the process of division. Gradually there were mutations (variation in genetics) and the new mutated form of cells began sexual reproduction. Living beings mutated into several variants, to adapt to the environmental variations, during the initial phases of the Earth formation. Today we see people of different temperate zones having different skin tones and other features. So continually the universe is experimenting and evolving newer types of matter, but without breaking its rules, which is evident in repeated trials carried out by scientists. This is the way in which the universe is unfolding, but surely if there was anything written, then it would not be this dragging. Also, our perception of time may only be imaginary as we are affected due to change in time. Hence the final conclusion I would like to draw here is that with the existing laws of nature, the universe is trying to evolve and each time it tries to produce something new, therefore it is illogical in assuming that our destinies are already chalked out.

Sreedhar Iyer

The pale blue dot

In the year 1990, Voyager 1 was about 6.4 billion kilometers (4 billion miles) away, and approximately 32 degrees above the ecliptic plane, when it captured this portrait of our world. Caught in the center of scattered light rays (a result of taking the picture so close to the Sun), Earth appears as a tiny point of light, a crescent only 0.12 pixel in size.

Reproduced below are the famous words on our planet Earth by Inspiring science and astronomy writer Carl Sagan:

Look again at that dot. That's here. That's home. That's

us. On it everyone you love, everyone you know, everyone you ever heard of, every human being who ever was, lived out their lives. The aggregate of our joy and suffering, thousands of confident religions, ideologies, and economic doctrines, every hunter and forager, every hero and coward, every creator and destroyer of civilization, every king and peasant, every young couple in love, every mother and father, hopeful child, inventor and explorer, every teacher of morals, every corrupt politician, every "superstar," every "supreme leader," every saint and sinner in the history of our species lived there--on a mote of dust suspended in a sunbeam.

The Earth is a very small stage in a vast cosmic arena. Think of the rivers of blood spilled by all those generals and emperors so that, in glory and triumph, they could become the momentary masters of a fraction of a dot. Think of the endless cruelties visited by the inhabitants of one corner of this pixel on the scarcely distinguishable inhabitants of some other corner, how frequent their misunderstandings, how eager they are to kill one another, how fervent their hatreds.

Our posturing, our imagined self-importance, the delusion that we have some privileged position in the Universe, are challenged by this point of pale light. Our planet is a lonely speck in the great enveloping cosmic dark. In our obscurity, in all this vastness, there is no hint that help will come from elsewhere to save us from ourselves.

The Earth is the only world known so far to harbor life. There is nowhere else, at least in the near future, to which our species could migrate. Visit, yes. Settle, not yet. Like it or not, for the moment the Earth is where we make our stand.

It has been said that astronomy is a humbling and

character-building experience. There is perhaps no better demonstration of the folly of human conceits than this distant image of our tiny world. To me, it underscores our responsibility to deal more kindly with one another, and to preserve and cherish the pale blue dot, the only home we've ever known.

-- Carl Sagan, Pale Blue Dot, 1994

Sreedhar Iyer

Conclusion

Botanical, archaeological and astronomical evidence suggest that we as a life form, made up of atoms and molecules, capable of converting energy from one form to another, contribute a small part in balancing the complex ecology of our planet. We are in fact complex supersets of cells which themselves are a combination of complex molecules. We are endowed with imagination which is possibly divergent from those of our ancestors.

Our technology has upgraded from agricultural implements to the Internet of Things and Artificial Intelligence, and we continue with the futuristic environment. From elusive sub-atomic particles, nature knitted the basic elements. Churning these elements with the help of gravity, it took several millions of years to make elements of higher order, and then it created allotropes and compounds.

The emergence of life on Earth has taken the creation to a new and complex level. Presently human beings are the mass survivors and seem to be the intelligent beings in the absence of any alien intelligence. Humans have taken on nature's work by creating experimentation labs like the CERN as a result of which we may advance the complexities of the universe. Creation of artificial intelligence is another area where we may be taking the universe's complexity to the next level.

With impending scenarios like global warming or meteor strikes, the Earth may become uninhabitable and selected few may attempt to step out of our humble planet, or out of our solar system to continue the journey of life elsewhere. There we may mutate further, adapting to new environments and planetary systems. Here, if you ask whether we should still continue in our quest on the origin of universe and life, then do not brood over it. Some of us are already mutated enough to keep this process going on.

Science does not directly accept God, as it solely believes in proof and experimentation. To be accepted by science requires that the theory is provable any number of times. Hard convincing evidence to prove the existence of God as depicted by religious organizations does not exist and moreover God can neither be seen by naked eyes nor felt through any means. Random incidents of supernatural occurrences cannot be repeated and hence science does not fully accept the concept of God, at least the way it is projected.

Some scientists hypothesize that universe was created by the big bang and also talk about possibly a big crunch. Also, there are theories of parallel universe or multiverse which are still in theoretical stage and merely a hypothesis. Science still does not have a provable theory on the creation of the universe, and the big bang theory also is only a hypothesis. The next mystery is that of the creation of life which is nowhere common. How the formation of life was triggered and how it sustained remains unanswered.

Religion says that God created the Earth, the sky,

and the stars, and we believed it. The spiritualists said that we are made of body and soul, and we believed it. Science once said that the smallest indivisible part of matter is atom and we believed it. Science later clarified that atoms are made of sub-atomic particles and that the electrons revolved around the central nucleus consisting of protons and neutrons and we accepted it. Further science again corrected and said that sub-atomic particles are not the basic building blocks of matter, but are made out of quarks and the leptons. Now we accept this theory too. Science said that the electron is either a wave or matter and changes its form notoriously, and we have accepted it.

Many of us believe whatever is put forth to us without any forethought. Only a few of us do not take things granted and try to reason it out or try to find a suitable explanation to the happenings around. Religion personifies God and gives God a human figure. Some religions believe a human leader as an incarnation of the supreme God. Christianity believes Jesus to be an incarnation of God. Hindus believe Brahma, Vishnu, and Maheshwara as the basic Gods and Rama and Krishna as the incarnation of Vishnu. Spirituality seeks to experience the unseen and tries to seek the ultimate reality or the eternal bliss. Science still does not have any definite theory for creation and existence. Science estimates that it can see only 5 percent of the universe and is unable to detect the remaining 95 percent of what exists in the universe.

Whatever we humans theorize or argue, our universe itself is a gigantic brain possibly several times more intelligent than our humble brains, creating and destroying galaxies, stars, planets and complex structures of the ecology of which we are a part of. Whether we will be able to understand the giant mechanism with our limited observational skills or, is there a bigger surprise beyond the visible universe or, are we just a part of a big dream or an illusion? Will we decode the secrets of the universe or continue to exist with uncertainty in this confusing, perplexing, enigmatic expanse?

The Enigmatic Expanse - Existence

ABOUT THE AUTHOR

Sreedhar Iyer was born and grew up in Bangalore. He now resides in Karwar, India. He is an avid lover of astronomy and is deeply interested in the universe, stars and galaxies. He always wonders the way universe works, the connection between human and nature and the purpose of life.

Starkly different views of theology, spirituality and scientific theories proposed by scientist such as Stephen Hawking and Carl Sagan have inspired him to write "The Enigmatic Expanse (Existence)."

CONNECT WITH THE AUTHOR

Email: sreedharpiyer@yahoo.com
Facebook: https://www.facebook.com/sreedharPiyer
Twitter: https://twitter.com/sreedharIyer
Instagram: https://www.instagram.com/psreedhar2000

www.ingramcontent.com/pod-product-compliance
Lightning Source LLC
Chambersburg PA
CBHW052331220526
45472CB00001B/378